Emmanuel odoyi

Challenges and Opportunities of Sustainable Electricity Access in Nigeria

GRIN Verlag

Bibliografische Information der Deutschen Nationalbibliothek:

Die Deutsche Bibliothek verzeichnet diese Publikation in der Deutschen National-
bibliografie; detaillierte bibliografische Daten sind im Internet über http://dnb.d-
nb.de/ abrufbar.

Imprint:

Copyright © 2013 GRIN Verlag GmbH
Druck und Bindung: Books on Demand GmbH, Norderstedt Germany
ISBN: 978-3-656-64187-2

Challenges and Opportunities of Sustainable Electricity Access in Nigeria

Sustainable Development

By

Emmanuel Joseph Odoyi

Abstract

Electricity is an important feature for economic development. The necessity lies in the fact that electricity affects every aspect of our economy. Nigeria should aim towards sustainable electricity access to its citizens. Nigeria must have a stable electrical energy supply to meet the market, basic needs of the people. A nation whose energy supply is not reliable does not help in the development of the national economy. Nigeria is a nation with more than 162 million inhabitants with less electric access to both rural and urban areas.

This study has reviewed and analyzed the challenges and opportunities facing Electricity Access in Nigeria. The study has employed approach to good features of sustainable development system in the Nigerian electrical energy sector by proposing a sustainable strategies and alternatives for energy access through the use of solar power, wind and hydro power, and Bio-mass . Also, the study clearly explained and analyzed the problems facing the electricity in Nigeria. Thus, this study has used literature review methods, to gather information's on Energy Sector in general and existing literatures on the Nigeria Energy Sector as well as a SWOT/PEST Matrix of the Nigerian Energy sector.

Keywords; Electricity access, energy poverty, energy efficiency,
sustainable renewable energy

Table of Contents

1 INTRODUCTION

Nigeria is a nation in west Africa with various energy resources, and abundant sources of energy like crude oil natural gas , coal and it is also endowed with large renewable energy resources such as, wind power, hydro power, bio-mass and solar power.

1.1 Research Area

Nigeria is a nation with enormous oil wealth, but many of its citizens do not have access to effective electricity supply. According to (Babatunde & Shuaibu 2009, p1), Nigeria has approximately 5,900 megawatts (MW) of installed electric generating capacity. The Power outages are frequent and the power sector operates well below its estimated capacity. A fundamental reason offered is the low generating capacity of the Nigerian power sector relative to installed capacity. Even with observations and research in Nigeria Electricity Access has become one of the country´s most significant problems. As electricity use is seen as the engine that drives the nations economy. But, because lack of Electricity access to every household, the national governments and the policy makers want to ensure that the citizens of Nigeria has adequate access to electricity both in rural and urban areas.

This is because, full electricity development in Nigeria will drive and enhance the national economy, because electricity is seen as the engine that drives industrialization in every nation's economy, as electricity access has many important benefits and potentials; 1) a stable Electricity access supply can become a key for the Nigerian nation to becomes one of the most developed in the world especially in terms of economy and the marketing environment as well improving the communication system to helps innovation in science and technology and providing adequate healthcare delivery system and improving the citizen's standard of living. According to (Emovon and Kareem 2008,) it's unfortunate that the biggest problem in Nigeria is electricity crisis, a crisis without end. Despite all these inherent problems associated with Electricity access in Nigeria, there is a growing call for the need of a significant reform that will transform the energy sector in general. This study tends to fill in the gap in understanding the challenges and the opportunities affecting the Electricity access in Nigeria

1.2 Research

One of the United Nations Development Program (UNDP) goals across the world is to strengthen nations in their ability to manage their resources in a sustainable way. The Energy Sector in every national government has significant impact in the lives of every citizen in every nation; the role of every government is to provide the basic needs of its citizens in ensuring a great and significant effect in a way to improve economic development. The need for electricity is paramount for the growth of a country, access to electricity as the basic form of energy supply to the masses is vital for the development of a nation's economy because of this; Electricity access is particularly crucial to human development as electricity is, in practice, indispensable for certain basic activities, such as lighting, refrigeration and the running of household appliances, which cannot easily be replaced by other forms of energy. Every individuals' access to electricity is one of the most understandable and un-distorted indication of a country's energy poverty status. This study will explore most of the challenges and the opportunities of having a viable electricity access in Nigeria and proposing strategies to help Nigeria in providing its citizens with clean and efficient energy supply.

1.3 Research Method

This study will review and analyze the challenges and opportunities facing the Electricity Access in Nigeria, the study will employ features and approach to good Sustainability that will clearly demonstrate how to understanding of the challenges and the opportunities in accessing electricity in Nigeria. This study will use the theoretical research methods, the theoretical will be beneficial to the study because, it will be useful to gather literature on Energy Sector in general and existing literatures on the Nigeria Energy Sector.

This study tends to fill in the gap in understanding the challenges and the opportunities affecting the Electricity access in Nigeria and proposing strategies to help Nigeria in providing its citizens with clean and efficient energy supply.

Also, PEST and SWOT Analysis will be used in other to understanding the challenges and opportunities of the Nigerian Electricity Access System.

2 ACCESS TO ELECTRICITY IN NIGERIA

Access to clean and sustainable electricity has become a major problem facing Nigeria, Nigeria is a nation with vast majority of abundant energy potential, still the country's energy crisis has challenged its ability to reduce poverty and support for free market economy as well as strengthening the socio-economic development of the country. For over two decades, Nigeria has experienced problem in electricity generation, transmission and distribution making electricity inaccessible. According to the IAEA report of 2011, it estimated that in 2009 about 1.3 billion people in the world lacked access to electrical energy of which 590 million lived in sub-Sahara Africa, especially in the rural areas. Presently, access to electricity in Nigeria is through power holding company of Nigeria (PHCN) which its power generation capacity is estimated to be around 6,000megawatts, with average working capacity of 2000 megawatts, to provide electricity to over 150 million people, this result to high importation of electricity generating sets (Generators) in other to have alternative electricity supply access

However, the inability of the governments to provide better electricity access to its citizens has made the nation; to fall short of achieving the millennium development goal. For instance, the ability for many great nations being able to achieve better basic education was due to the fact that their educational facilities has the required necessary electricity facilities as a basic teaching tool for their citizens ability to read and write, this is because basic lighting for home may help reduce child mortality and improvement of maternal health. Also, electricity is needed for better health facilities as to enable preservation of drugs and vaccines in the refrigerator as well as better service delivery for emergency and intensive care to patients in the hospitals, Electricity energy is needed for better health living, cooking of foods for a comfortable living temperature, lightings for appliances, piped water or sewerage, educational aids, communication technologies such as radio, televisions, emails, and the internet) and transport, Energy also Fuels productive activities such as farming, trading, industrialization of products and mining activities and leisure and sports.

Similarly, lack of electrical energy access can contribute to energy poverty, and deprivation can lead to economic decline of a nation. Even though, energy and poverty are not fully integrated together but with the socioeconomic development which involves productivity, income growth level, health and education Ajayi and Ajanaku (2007).

2.1 Nigeria and General Statistic

Nigeria is situated in Western part Africa. It bordered Niger and Chad through the north, and Cameroon through the south, and Benin through the west, it's also surrounded by the Gulf of Guinea which borders Togo and Ghana by the water.

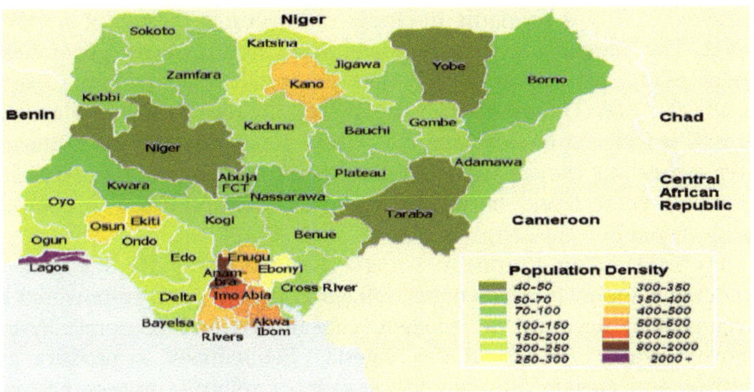

Figure 1. Map of Nigeria[1]

Table 1. Statistic

Capital(s)	Abuja
States:	36 States and Capital
Population:	140,431,790 (2006 Census)
Area:	923,768 km^2
Currency:	₦158.80 Naira = 1 US Dollar
Language(s):	English, Hausa, Igbo, Fulani, Yoruba and Ijaw
Time Zone:	GMT+1h00
ISO Code:	NG
Dialing Code:	+234
Internet Tld	.ng

---

3   INTRODUCTION OF PHCN (POWER HOLDING COMPANY OF NIGERIA)

Power Holding company of Nigeria (PHCN) is owned by Nigeria Government. It is responsible for electricity production and supply in the country. It is active in the area of generation, transmission and distribution of electricity. PHCN is primarily in charge of governing the use of electricity in Nigeria. The company generates power from hydro power generation and thermal plants generation. PHCN has an installed generating capacity of around 6,000 MW. However, the company usually records low availability. The company is focusing on development of new generation projects in the area of power generation, transmission and distribution. PHCN was formed in 2005 as a result of vertical and horizontal unbundling of NEPA. The company is headquartered in Abuja, Nigeria. (Global Data 2011)

Power Holding Company Nigeria has monopoly for the generation, transmission and distribution of centralized grid power and empowered by the government to keep up an efficient, coordinated and economic system of electricity supply throughout the Nigeria. Sequentially to produce greater efficiency and sustainability in the system, a reforms process has was in progress to remove government's monopoly on generation, transmission and distribution. Thus, in 2005, the Electric Power Sector Reform Act (EPSRA) was enacted. The law is intended at liberalizing the power sector.

The PHCN started with four major power stations namely: Ijora, Delta, Afam thermal power station and Kanji hydro power station, providing electricity for about five million customers nationwide, In turn increases industrial growth. As a result of the reform process, the distribution and generation areas of the electricity market are deregulated. PHCN is grouped into six different generation companies and eleven distribution companies for the various parts of the country.

According to the (Central Bank of Nigeria, Research Department) PHCN currently accounts for about 98% of the total electricity generation. Power generation by other agencies such as the Nigerian Electricity Supply Company relies on thermal power for electricity generation unlike PHCN, which relies on both hydro- and thermal power. However, electricity is also a consumer of fuel and energy such as fuel oil, natural gas, and diesel oil. The importance of these sources of energy and fuel for generating electricity has been decreasing in recent years. However, hydropower that is relatively cheaper than these sources has grown to be more important than other sources (Famuyide, et al., 2004). Thus, more recently, the Power Authority has generated electricity through a mix of both thermal and hydro systems.

The national electricity grid presently consists of 14 generating stations (3 hydro and 11 thermal) with a total installed capacity of about 8,039 MW. The transmission network is made up of 5,000 km of 330-kV lines, 6,000 km of 132-

kV lines, 23 of 330/132-kV substations, with a combined capacity of 6,000 or 4,600 MVA at a utilization factor of 80%. In turn, the 91 of 132/33-kV substations have a combined capacity of 7,800 or 5,800 MVA at a utilization factor of 75%. The distribution sector is comprised of 23,753 km of 33-kV lines, 19,226 km of 11-kV lines, and 679 of 33/ 11-kV substations. There are also 1,790 distribution transformers and 680 injection substations Fagbenle et al., (2006).

Despite the current power situation in Nigeria, and in spite of the current deregulation process, the Federal Government is still increasing investments in the sector to increase the generation capacity to 10,000MW by 2011 and to 20,000MW by 2015 through the National Integrated Power Projects being implemented across the country. Most of the additional capacity would be thermal (Gas) plants with renewable such as solar, biomass and wind contributing a minimal amount. Nuclear power has been factored into the energy mix, and it is expected that the first nuclear power plant of 1,000MW capacity is expected to be commissioned not later than 2020 (Erepamo, 2009).

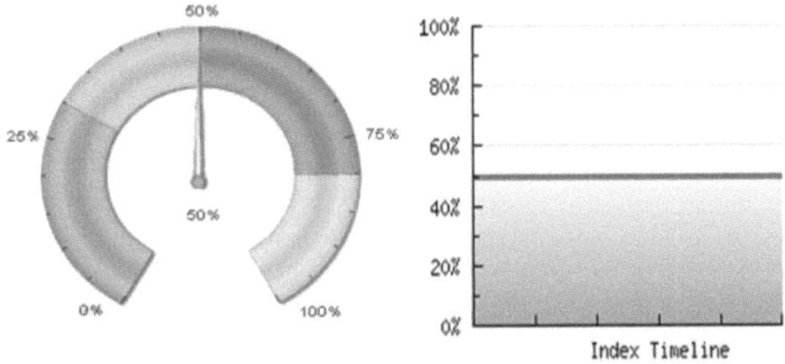

Figure 2: Index Score (New statesman 2013)

- Power Holding Company of Nigeria Currently Scores 50% in the index.
- Power Holding Company of Nigeria Currently ranked equal 15 out of 35419 included in the NS Company Index. This is the top of 0.042% of all companies.
- Power Holding Company of Nigeria is currently ranked 3 out of 10785. This is the top 0.028% percent of Energy companies ranked in the index.
- User perception of the company stands at 0% percent. This differ 50% percent over the score attributed to the company by other scoring factors. presently PHCN comprises of the following:

Three hydro and seven thermal generating stations with installed capacity of about 6,852MW and available capacity of 3,542MW , a radial transmission grid (330KV and 132KV) and Eleven distribution companies(33KV and below) in charge of wires, sales, billing, collection and customer cares (PHCN, 2013). However, the Nigeria government and the 14 preferred Bidders (Company) created out of PHCN as a result of the reform process has sign a share sale Agreement and Concession Agreement. The government and the Preferred Bidders will work together to archive a successful facilities handover to the Bidders, these has been regarded as a boost and in turn will increase generation, transmission and distribution capacity and will increase sustainable electricity access in Nigeria.

4 CHALLENGES OF ELECTRICITY ACCESS IN NIGERIA

Despite diversification of energy sources for domestic consumption, energy price reform has not been effectively pursued and energy intensity remains high, posing a serious threat to the economy. In most urban and rural areas in Nigeria, electricity installation and consumptions are very substandard. This is because; the present urban-centered energy policy is very poor, as cases of rural and sub-rural energy demand and supply do not reach the centre stage of the country's energy growth plan. People in rural areas depend on burning fire wood and traditional biomass for their energy needs, causing great deforestation to the people by loss of solid soil as well as discharge green house gases and pollution of the environment, global warming and other environmental problems.

Thus, the main responsibility of the government has been to supply electricity to the metropolis, and other different places of industrial developments, therefore, creating energy disparity within the country's socio economic and political environment. With the present measure and the ever changing increase in the population with total capacity of the available power stations reveals that Nigeria is not able to meet its energy need of its people; the rural residents still lack electricity access. These people have resorted to the most popular and alternative source of energy dependant with the use of generators.

According to a recent report by the Vanguard New paper April 2013, Nigerian has one of the most problematic electricity sectors in the world, with an estimated installed electricity generation capacity of 8,644 MW, and available capacity of only approximately 3,718 MW, to cater for the needs of a population of over 160 million. Even as, the Nigerian government has recently announced the commencement of the review of its "Power sector Roadmap" in other to tackle the reform process especially in the areas of granting of distribution licenses, capacity shortages and, generally letting the bottom fall out of the quality of power generation and distribution. Still, these reforms have not been achieved. Though, many of the challenges in the electricity sectors are;

4.1 Inadequate Infrastructure

The inability of the government to provide adequate infrastructure in the energy sectors has made the industry not to function effectively. Also, the energy sectors supply of system components are often dependant on importation overseas as well as modern technologies to monitor vandalisation, distribution, transmission of electricity. While, the technical expertises to develop, deploy and manage energy are inadequate.

4.2 Underutilized Energy /Human Resources

Under utilisation of energy alternatives sources (sustainable renewable energy).thus, all renewable resources for electricity are available in Nigeria, but they are not properly harnessed or. developed, thereby leading to lack of efficient electricity access.
Also, Nigeria has considerable human and material resources to modernize its energy supply and to co-ordinate the transition to a sustainable energy system. Besides, Nigeria has a high amount of renewable energy sources: which are in favourable conditions and more profitable especially with the use of wind energy, other opportunities are also for solar energy.

4.3 High Investment Cost

Another challenges facing the Nigerian power sector is the ability to generate high unlimited electricity access supply. The main obstacle is to increase the electricity generation capacity of which investment cost in electricity is high, and expensive in because of high cost of electricity production

4.4 Energy Mix

Energy mix is challenge facing the power sector in Nigeria. The electricity sector has been powered by hydro and thermal plants. These have not been sufficient to meet the electricity need of the country. Nigeria is blessed with abundant solar and wind energy, which are yet to be fully exploited in generating electricity Makwe (2011). According to Foster et al (2011), Nigerian tariff is one of the lowest in Africa.

5 ENERGY USE & ENERGY EFFICIENCY IN NIGERIA

According to Sustainable Energy watch (2007) the basic primary structure of energy usage in Nigeria is based on natural gas which is 43.3%, petroleum products 22.3%, hydropower 4.3% and traditional biomass 30.1%. While, energy demand by sector is distributed as follows; industry 10.7%, transport 29.9%, household 45.9% and services 13.5%. Also, traditional fuels make up about one third of primary energy consumption while modern renewable energy consumption represents less than 5% with energy anticipated development of coal power plants; the fossil content of the Nigerian energy consumption pattern is increasingly fossil based.

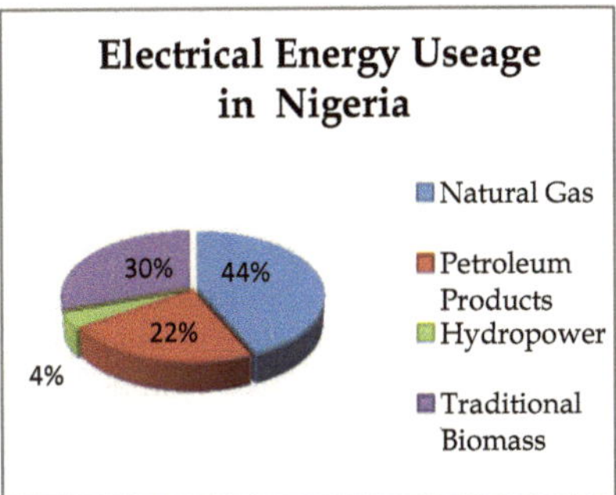

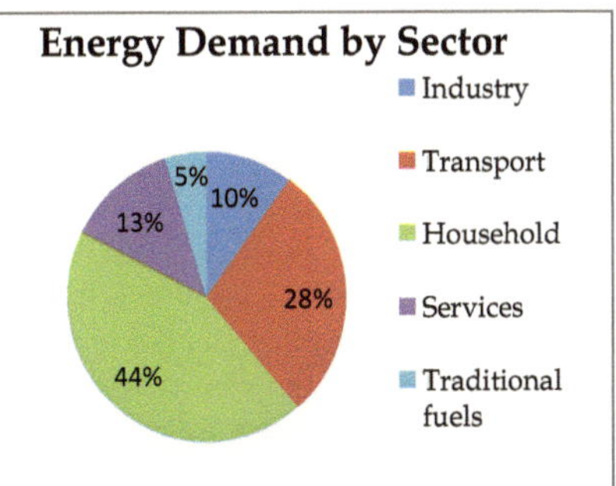

Figure 3: Electrical Energy Usage

Electrical energy use-ability patterns in the world today shows that some countries in Africa and India have the lowest rates of energy usage and consumption across the world. Nigeria is in this particular group, on the other hand, Nigeria suffers from an inadequate supply of electricity energy due to the rapid growing demand, which is typical of a developing economy. Ironically, Nigeria is potentially endowed with sustainable energy resource; because Nigeria is rich in traditional energy resources, which includes oil, natural gas, lignite, coal and other renewable energy resources such as wood, solar, hydropower and wind Energy commission of Nigeria (ECN) (2003).

The model of electricity usability in the Nigerian economy has been divided into many sub-sectors of the economy such as industrial, transport, commercial, agricultural and household sectors. The household sector accounts for the largest share of energy consumption in the entire country of about 65%. This is largely due to under development in all other sectors in electricity usage. While, the main forms of electricity consumption in the household always include cooking, lightings, and the use of appliances; cooking account for 91%, lightings 6% and the remaining 3% are for basic uses in electrical appliances such as pressing ironing and televisions Energy commission of Nigeria (ECN) (2005). The predominant energy resources for domestic and commercial uses in Nigeria are fuel wood, charcoal, and kerosene, cooking gas, and electricity Famuyide et al., p1-7, (2011). Other sources, which are less common, are saw dust, agricultural crop residues of corn stalk, cassava sticks and in extreme cases cow dung. In Nigeria, among the urban inhabitant, kerosene and gas are the major cooking fuels. The majority of the people rely on kerosene stoves for domestic cooking, while only a few use gas and electric cookers Abiodun (2003). The rural areas have little access to usual electrical energy and petroleum products due to the absence of good road networks. Therefore, petroleum products such as gasoline and kerosene are purchased in the rural areas.

5.1 Energy Efficiency

However, energy efficiency has been described by Diesendorf 2007, as an objective in other to reduce the amount of energy required to provide products and services, for example insulation of a home allows a building to use less heating and cooling energy to achieve and maintain a comfortable temperature. For instance, installing fluorescent lights or natural lights to reduce the amount of required energy to attain the same level of illumination compared with using traditional incandescent light bulbs, improvement in energy efficiency are often achieved by adopting a more efficient technology or process of production.

There are numerous reasons to improve energy efficiency, because reducing energy use reduced energy costs and may result in a financial saving to consumers. Also reducing energy use is seen as a solution to the problem of reducing emissions, because according to IEA, improved energy efficiency in buildings, industrial processes and transportation could help reduce world's energy needs in 2050 by one third and also help control the world global emission of green house gases Hebden (2007).

Though, energy in Nigeria is still having the challenges of energy inefficiency, the power holding company of Nigeria have not been able to give adequate electricity to the population of Nigeria. The energy sector at large has a lot of challenges in meeting the required demand for electrical energy.

On the other hand, countries like Nigeria has not been able to see an opportunity through the use of its energy efficiency, it produces lots of energy, but it population are not able to benefit from these options, because the government of Nigerian exports a lot of energy and leave its people without efficient energy. The chart below show countries Nigeria is supplying Liquefied Natural Gas across the world.

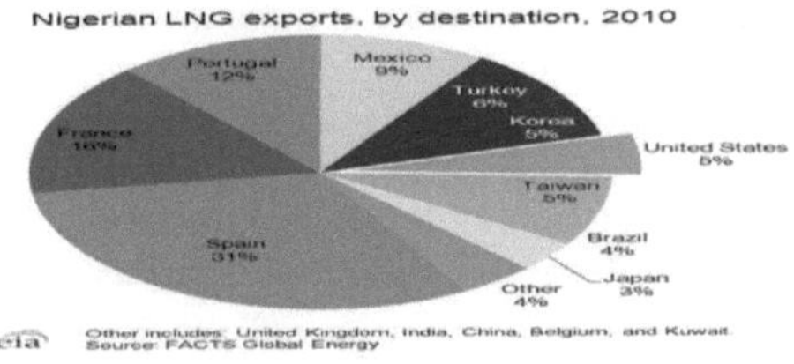

Figure 4: Nigerian Gas Export

However, Energy Improvement is an essential factor of household's demand for electricity. Individual households demand for electricity is a derived demand, it is derived from the demand for services such as, heating, and cooling obtained from using appliances and equipment (examples as heaters and air conditioners). However, utility obtained from using electrical equipment and appliances depends on the technology embedded in such equipment and appliances. Since individual households are tied to fixed equipment and appliances, they are constrained in the services they obtain from using appliances, however, in the long run due to progress in knowledge such as the learning of new demand management including timing appliances for switching thermostats, etc on and off this will bring about an appreciable increase in energy efficiency. Furthermore, the modification of existing appliances and introduction of new ones in to the market will change the technical characteristics of the appliances and equipments which can also bring about efficiency, thereby reducing overall electricity consumption for a long-term usability Uyigue et al., (2009).

5.2 Electricity Access and Improving access

Access to sustainable, affordable and modern electricity is a basic requirement for poverty reduction and sustainable human development. Electricity access in Nigeria is through PHCN and generator. Presently, 15.3million households lack access to grid electricity and for those connected to the national grid, supply is erratic at best. Lack of electricity is beyond access to electricity; about 72% of Nigerians depend on wood as source of fuel for cooking, ironing, and lightings.

Every individuals' access to electricity is one of the most understandable and un-distorted indication of a country's energy poverty status. In other to access efficient energy and improve the access to energy in Nigeria, the government needs to improve their own energy data survey. Also, there are opportunities of making good use of off-grid generation in other to improve access, and looking at alternative energy options in other to achieve sufficient access.

However, several reports have highlighted many common barriers to electricity access especially by poor urban dwellers and communities in most developing nations were high cost of services, this is because the poor often face multiple constraints especially in the areas of disposable incomes due to the nature of their informal livelihood. In other words, they are often not able to afford the infrastructure costs such as meters, wires, appropriate stoves, and safe construction materials. Often they face the challenges of untimely bill payments which often result in disconnection of services. Also, illiteracy often makes people in some certain areas as slums, or local communities are not often aware of the financial benefit of clean and effective electricity access. Sometimes, lack of mutual trust and co-operation between local communities and energy service providers to reach an understanding can prevent good relationship for a better electrical access and sometimes lack of proper and adequate infrastructure Arora et al, (2011).

However, in other to achieve a better and a sustainable electricity access, the governments should be able to provide subsidies for the poor urban and local communities, and creation of better employment of labour in these areas in other for these people to access adequate financial status. Also, the communities or slum dwellers can have a memorandum of understanding to be able to have a relationship and trust. And finally, the government should provide adequate housing for its citizens to be able to have a standard of electricity access.

6   ENERGY POVERTY IN NIGERIA

The availability of Energy (electricity) is one of the key to industrialization and socio-economic development of any economy when effectively harness, develop and utilize. Energy availability can be seen as a factor in attaining Millennium Development Goals but in Nigeria the reserves is the case since there exist Energy poverty in Nigeria. Energy poverty is lack of energy or limited access to energy (electricity). In Nigeria despite the abundant natural resources in Nigeria e.g. wind, solar, coal and hydro, still there is energy poverty.   About 85 million Nigerian which is 60percent of Nigeria population have no access to electricity or limited access to electricity service. Electricity consumption per capita is about 100Kwh, however because of the energy poverty that affects both the rich and poor in Nigeria, Nigerians have resulted in providing alternatives e.g. household resort to generators for electricity supply.

6.1 Impact and Implication

Lack of electricity intensify poverty, it can hinder job creations and most industrial activities leading to unemployment, most company whose business depend on electricity try to manage the energy poverty are faced with high running cost and are forced to lay off workers resulting in unemployment. Energy poverty is a perennial social problem affecting most developing countries not just Nigeria alone. Statistics shows that 1.6 billion people have no access to electricity, 80% of them in south Asia and Sub -Saharan Africa. Four out of five people without electricity live in rural areas of the developing countries. 2.4 billion people in the world lack modern fuels or rely on traditional bio-mass wood, agricultural residues and dung for cooking and heating (Club De Madrid, 2007; IEA,2002). According to Club De Madrid (2007) states that energy poverty is a major barrier to growth and development in vast areas of the world. Meaning countries that want to develop must address energy poverty. Energy poverty contributes to the poor health and premature death.

According to World Health Organization (who) estimates that 2.5 million women and young children in developing countries die prematurely each year from breathing the fumes from indoor biomass stoves. In household in the urban and rural areas uses stoves and generators that emit carbon monoxide and various toxic gases that causes respiratory illness.

Energy poverty also affect the academic institutions(Students and lecturers) because lack of regular power supply will limit or increases hours spent in studying, sourcing of information on the web, research and other teaching activities e.g. using projectors or modern teaching facilities.

6.2 Environmental Impacts

The energy industry in Nigeria has a harsh environmental consequence, which are mostly in forms of both pollution and deforestation Rilwanua (2003). Ultimately, the most imminent energy issues in Nigeria are not related to the environment, but to social welfare. Although, the immediate environmental consequences of current practices are not on the scale of the current social needs irresponsible current environmental practices now could translate into disastrous impact in the future. NEPA which is now called PHCN is not considering the environment as its main priority, but it has promised to support energy sector reform through environmental friendly process.

The main contributors to air pollution are mainly through gas flares, as well land and water pollution in the Niger delta region as the oil firms and the governments are not talking full responsibilities of major environmental concern. However, the government has pledged to stop the gas flares in 2008 but, presently, the gas flares have been tuned to smokeless gas flares. However, below are outlines of the government's strategy to proceed with energy development through environmental friendly method;

(i) Putting adequate standard in place through strengthening the regulatory agencies and developing definitive goals that must be met and (ii) Assess the environmental impact of energy projects and providing alternative to fuel-wood and encouraging research and development Rilwanua (2003).

However, the federal ministry of environment has the general responsibility for the environment in Nigeria. The ministry had little power to enforce environmental laws, regulations and standards. Environmental governance has been strengthened by establishing the General National Environmental Standard and Regulations Enforcement Agency (NESREA) in 2007. However, NESREA is responsible for enforcing all national environmental laws policies and regulations, and also for enforcing fulfillment with international conventions. NESREA has 12 state offices and 4 zonal offices. In addition, states and local government council have been encouraged to set up their own environment protection agencies. The federal ministry of environment has also a special climate change unit. The sustainable use of natural resources and environmental protection has been recognized as priorities in country's economic growth and modernization plans. However, the environmental minister has also noted various environmental challenges which were mainly in the areas water pollution, indoor and outdoor pollution, industrial pollution, biodiversity loss, flooding, land degradation desertification, climate change. Although, these environmental problems has been directly linked to failures in adhering to the already established rules and regulations of the environments (Special Climate Change Unit News, August 21st 2011).

7 OPPORTUNITIES

According to a recent report by the IAEA has projected the need for investment in the world electricity industry because of future electricity supply demand of 1.7% per year around the globe. The investment needed in this industry is projected to be around amount to $16 trillion, or $550 billion a year, which will be measured from 2001-2030. Which the capital needed will steadily grow through the projection period with the average of investment to rise from around $450 billion in the current decade to $630 billion in 2021-2030.

However, GE (General Electric) has promised to invest $1 Billion dollars in the Nigerian electricity industry especially through partnership with other Nigerian private firms in building major power plants, turbines, a new factory and exploring Nigeria's abundant natural oil and gas supplies Heather-Murdock (2013). In a nutshell, transforming the power sector in Nigeria will help create jobs and help the economy of Nigeria in general. Below are some of the opportunities for the future of the Nigerian electricity industry.

7.1 Investment Option as Opportunities

According to Okoro and Chikuni (2007) investing in the electrical industry is an opportunity for a massive expansion of personal share of ownership in Nigeria. It believed that over 800,000 investors can be created after the privatization of the power sector. Which, the government has welcomed the development as new step to a greater development that will enable economic growth and good energy future. However, it can reduce the reliance of public enterprise on the government for finance. However, it is good to privatized the power sector , and invest on it, especially, successive firms can easily raise funds through the capital market once the necessary investors confidence has been developed; thus this will enable them to change their growth and expansion of their business strategy. Also, new power facilities will provide the private sector new capital finance that will be injected into the economy, which on the long run will create future jobs in the industry. Power sector reform will create an enabling environment for investment and a healthy co-operative industrial setup. Also research and development are needed in the areas of services and innovative technologies to enable a brighter future in the investment areas. Also, investor can be invited by various sector of the economy, such as Nigerian banks, Federal government, Nigerian oil firms as well as foreign investors across the globe.

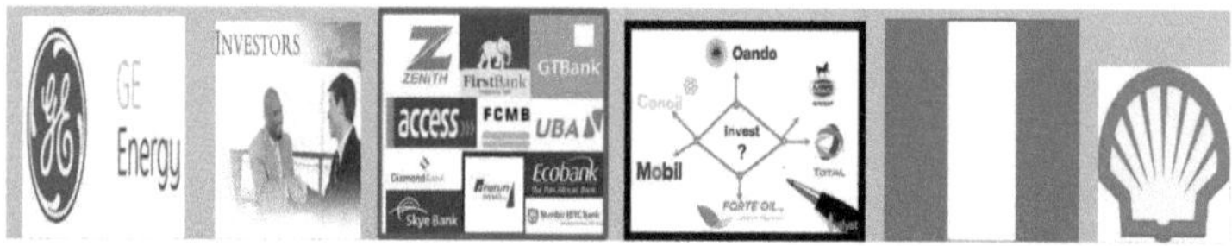

Figure 9: Potential Investors

7.2 Sustainability Electricity by Reform

However, a recent article published by the Business-day News march 2013 said that for Nigeria to make a substantial progress in the power sector, which Nigeria needs about $25bn investment in gas for a sustainable power reform, between now and 2020. According to "an oil and gas expert", the investment in the oil and gas are said should be channeled to infrastructure development across the gas power value chain.

Also, it is recorded that about $10 billion would be needed to develop gas fields, procession, and transportation networks. Another $12 billion would be required for power generation and $3 billion for transmission and distribution.

Though, an estimated $60 billion investment would be required, if the country is to attain 40gigawatts (40,000 megawatts) of electricity in the nearest future, with its current power production levels at 4,300 megawatts. Also, Mr. James Olotu, the Managing Director of the (Niger Delta Power Holding Company) limited has assured Nigerians that most of the projects in the country will be completed by December (2013).

7.3 Employment Opportunities

One of the opportunities of electricity access in Nigeria will create employment opportunities both in rural and urban areas of Nigeria. Because companies that depend on electricity for the production or manufacturing business will require skilled and unskilled labour in other to meet their business demand. Thus, these could also result in higher income level among the poor and middle income person.

7.4 Efficiency and Reliable Services

Proper availability of electricity supply will promote efficiency and growth in the power sector, and the Nigerian economy at large. These will lead improve electricity supply services, and will encourage foreign investors and domestic investors to invest in the electricity industry and other sectors of the economy. Thus, this will lead to efficient and reliable electricity supply.

7.5 Energy Alternatives and Strategies (Sustainable Renewable Energy)

Sustainable renewable energy is the alternative to sustainable electricity access in Nigeria, there is huge renewable energy sources like: Bio mass, wind power, solar power, Hydro power etc.

7.5.1 Biomass Energy: Biomass Energy is one of the renewable energy sources; it is the conversion of stored energy in plant to generate electricity or to produce heat from them. Biomass source are used and available in Nigeria whether urban or rural area are fuel wood (wood chips, dead trees), biodegradable and animals & plants used for fibres and chemicals production. Biomass energy systems begin by trapping sunlight in plants and production of chemical compound. it is called photosynthesis. Photosynthesis usually continues with efficiencies of less than 8%. In the biomass production process materials are usually transported from one site another by what is known as sap and the fruits that are involve in the process performs the sexual reproduction(DaRosa, 2008,p.g 569).

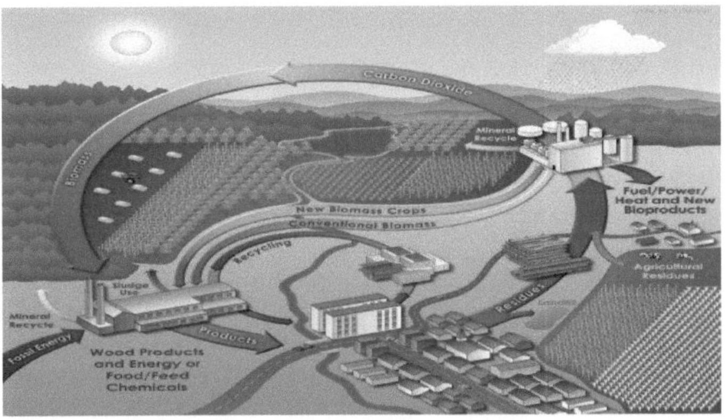

Figure 5: Generating Energy from Biomass (REPP, 2009).

In Nigeria, biomass resources have been estimated to be about 8×102MJ; plant biomass can also be utilized as fuel for small scale industries.
Biomass and petroleum fuel wood is over 90% of domestic energy need of about 70%of Nigerian population, for sustainable electricity access this system should be effectively develop.

7.5.2 Wind Energy: Wind energy is the conversions of wind for electricity generation (power) e.g. wind mills, wind pumps and wind turbine. The use of wind power for electricity will increase energy base which will reduce environmental pollution. The amount of the economically extractable power that is available from the wind is said to be more than the current human power use from all sources (Sorenson, 2008). In Nigeria, Wind energy is available at annual average speeds of about 2.0m/s at the coastal region and 4.0m/s at the far northern region of the country. With an air density of 1.1kg/m, the wind intensity perpendicular to the wind direction ranges between 4.4W/m, at the coastal areas and 35.2W/m, at the far northern region (Sambo 2009).

Figure 6: Wind turbine (W.Christian, 2006).

Wind power can be a sustainable renewable energy alternative for the provision of electricity supply for sustainable electricity access in Nigeria .The wind energy should be effectively develops in other to enable electricity accessibility to Nigerians.

7.5.3 Solar Power: Uses the energy that comes from the sun light to generate electricity. Nigerian is located in a high sunshine belt with huge solar energy sources. The solar radiation is fairly well distributed with average solar radiation of about 19.8MJm-2day-1 and average sunshine hours of 6hours per day. If solar collectors are use to cover 1% of Nigeria's land area, it is possible to generate 1850×10GWh of solar electricity per year ; this is over one hundred times the current grid electricity consumption level in the country(Uzoma et al., 2011).The generation of electricity from the sun can be by photovoltaic (pv) or concentrating solar power (CSP).

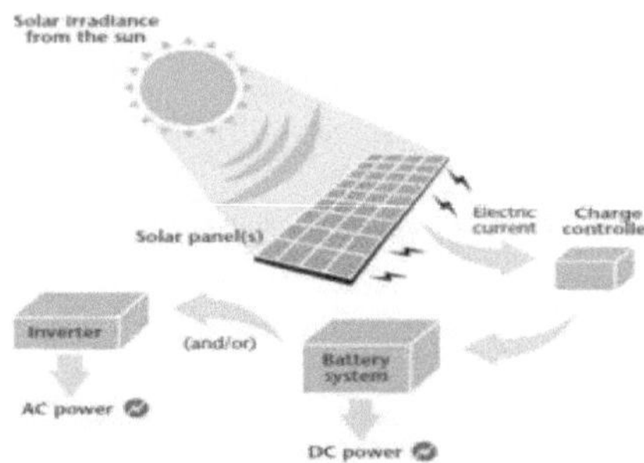

Figure 7: Solar-Power

Solar power can be a sustainable renewable energy alternative for the provision of electricity supply for sustainable electricity access in Nigeria .solar energy techniques should be effectively develops in other to enable electricity accessibility to Nigerians.

7.5.4 Hydro Power: Hydro power is the use of gravitation force when waterfalls or flows from an elevated or high place to a low place to generate electricity. Converting the flowing water to a usable energy produces hydro power (electricity). Nigeria have the potential to generate electricity through Hydro power, because of large rivers , dams and natural falls, also most streams and rivers exist and maintain minimum discharge throughout the year. The total technically exploitable hydropower potential based on Nigeria river system is estimated to be about 11,000MW of which only 19% is currently tapped or develop (Okafor et al; 2010).

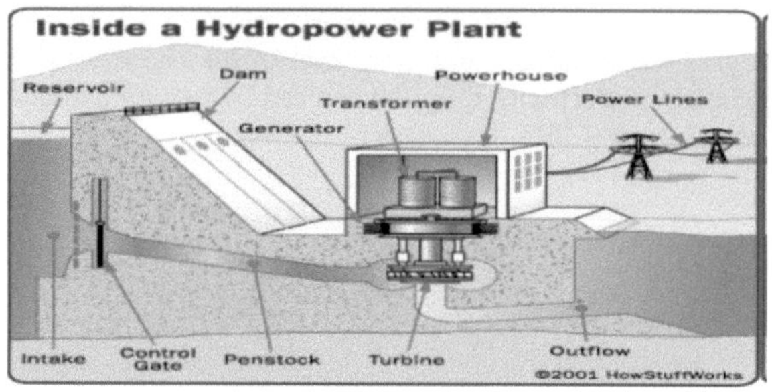

Figure 8: Generating Energy from Hydro (EMT, 2009).

Hydroelectric power stations have a turbine driven by the falling water from the dam. Hydro power can be a sustainable renewable energy alternative for the provision of electricity supply for sustainable electricity access in Nigeria. Hydropower techniques should be effectively develop in other to enable the provision of affordable and accessible electricity to Nigerians.

7.6 SWOT/PEST Matrix of the Nigeria Power Sector

	Political	Economic	Social	Technology
Strengths	Abundance of energy resources, Energy alternatives like; solar, wind etc.	Energy producing, high revenue from natural resources, investors	high in human capital, higher-educational institutions & multicultural society	computerization of energy data, private sector business & industrialization
Weakness	Federalism and energy Niger delta energy resource control, weak regulatory organizations	inadequate financing of the energy sector, inadequate data for planning, low taxation rate	energy -environmental issues, ethnic -conflicts Youth-restiveness, Infrastructure inadequacy.	lack of executive capacity in IT solutions in the energy sector, limited access to internet
Opportunities	democratic vs. undemocratic, strength of consumers on the Nigerian energy market, deregulation	Rising price of oil and energy, growing consumption for household electrical energy demand & supply, energy privatization and job creation	Cultural tourism potentials, developed market opportunities for good social system.	increased output of energy production cost savings, global energy plan, ecommerce, e-energy strategy
Threats	Power and energy sector corruption. electricity rationalizations, vandalisation and theft	overstretched energy finance, high cost of energy (decaying infrastructure), inappropriate managerial skills	energy poverty, environmental problems and unemployment	Negative values from the internet, IT security (hacking) Brain drain, loss of markets, increased threats of globalization.

8   CONCLUSIONS AND RECOMMENDATION

Nigeria has abundance of energy resources and alternative energy solutions which if developed fully, will satisfy the electricity needs of the country and enhance social, economic development of Nigeria. However, the government budget on electricity and investment should be implemented to its fullest. The sustainable energy alternative mentioned above should be used to improve electricity access to the country. The signed agreement between the government and the Preferred Bidders should be implemented in other to archive a successful facilities handover to the Bidders; this in-turn will increase generation, transmission and distribution capacity and will increase sustainable electricity access in Nigeria.

However, the adoption of renewable energy solution by the country with emphasis for development will lead to the internal reduction of petroleum products, gas and health risks emanating from the fume of the generators. The advantages of non-conventional energy solutions (renewable energy solutions) include ease of maintenance, simplicity of the technologies as well as the environmental friendliness over the conventional energy systems (fossil fuels).

However, looking at the energy crisis in Nigeria it is important that Nigeria state should be able to apply various technologies and managerial skills in other to help the energy industry, because this can in the future create a long-term sustainable development in the energy sector and the economy in general. The Nigerian society has not been able to sustain its energy needs and the crisis associated with it; especially the main industry to help in the energy sector is the oil and gas industries, which has massive tons of waste materials to convert to electricity, but the problem has continued the oil industries are involved with many cases of environmental pollutions and the same time the government is not taking adequate responsibility in penalizing some of the oil firms.

In other words, whether the Nigerian government is still planning to be able to achieve its energy goal, I therefore would suggest that the federal government of Nigeria should be able to re-visit some of the agreement it has signed with the oil companies in other for them to be able to manage their energy needs and also implement some of the legislations for adequate environmental protection of energy use as well more research and investment in these sectors.

REFERENCES

Abiodun R (2003). Fuel price Hike Spells Doom for Nigeria's Forest.
www.islamonline.net/English/index.shtml

Arora et al,. 2011. Improving Energy Access to the Urban Poor in Developing
Countries: The Energy Sector Management Assistance Program; The
World Bank November 2011

Ajayi & Ajanako (2007). Nigeria's Energy Challenge and Power Development:
The Way Forward. Bulletin of Science Association of Nigeria

Abideen A., 2011: The Power Sector and Industrial Development in Nigeria:
Lahti University of Applied Sciences: Degree programme in International
Business Thesis of Bachelor of Business Administration.

Anumaka M. C., 2012. Scenario of Electricity in Nigeria International Journal of
Engineering and Innovative Technology (IJEIT) Volume 1, Issue 6, June
2012 176

Baker L., 2008. Facilitating whose power? IFI influence in Nigeria's energy
sector: Published by Bretton Woods Project Hamlyn House, Macdonald
Road, London N19 5PG, UK

Babatunde M. A., & Shuaibu M. I., 2009. The Demand for Residential Electricity
in Nigeria: A Bound Testing Approach: Department of Economics Ibadan,
Nigeria

Bloomberg Financial Market
http://www.bloomberg.com/slideshow/2013-02-13/highest-cheapest-gas-
prices-by-country.html#slide55. Accessed 20.04.13

Central Bank of Nigeria (CBN) (2000) 'The changing structure of the Nigerian
economy and implications for development'. Research Department,
Central Bank of Nigeria; Realm Communications Ltd, Lagos, August

Connors S.R., 1998. Issues in Energy AND Sustainable Development
M.I.T. Energy Laboratory

Da Rosa et al., (2008). Fundamentals of Renewable Energy Process, 2nd edition.
Academic press.

Diesendorf, Mark (2007). Greenhouse Solutions with Sustainable Energy,
UNSW Press, p.86

Eleri, Ugwu & Onuvae (2012) Expanding Access to Pro-poor energy services in Nigeria: International Centre for Energy, Environment & Development http://www.iceednigeria.org/workspace/uploads/final-pro-poor-energy-access-paper-26-nov.pdf: Accessed 20.03.13

Eleri E.O., 2007. Strengthening Energy and Ecosystem Resilience in Nigeria: Sustainable Energy Watch 2007.
http://www.helio-international.org/Nigeria.En.pdf

Estes R. J., 2003. Toward Sustainable Development: From Theory to Praxis University of Pennsylvania, School of Social Work

Energy Commission of Nigeria (ECN) (2003) National Energy Policy. Federal Republic of Nigeria, Abuja

Ekpo U, N., Chuku A. C and Effiong E. L., 2011. The Dynamics of Electricity Demand and Comsumption in Nigeria: Application of the Bounds Testing Approach Department of Economics, University of Uyo, Uyo, Nigeria: Current Research Journal of Economic Theory 3(2): 43-52, 2011 ISSN: 2042-4841 © Maxwell Scientific Organization, 2011

Foster, V., and Pushak, N., (2011). Nigeria's Infrastructure: A Continental Perspective; Policy Research Working Paper, World Bank

Famuyide et, al., (2004) Socio-Economic Impacts of Deforestation in the Sudano-Sahelian Belt of Nigeria. Journal of Forestry Research and Management 1(1&2):94–106

Fagbenle et al., (2006) Draft Final Report on Nigeria's Electricity Sector Executive Report

Famuyide., Anamayi & Usman (2011). Energy Resources Pricing Policy: And it's Implications on Forestry and Environmental Policy Implementation in Nigeria. Continental Sustainable Development

Gallopín G., 2003. A systems approach to sustainability and sustainable development; Sustainable Development and Human Settlements Division; ECLAC/ Government of the Netherlands Project NET/00/063 "Sustainability Assessment in Latin America and the Caribbean"medio ambiente y desarrollo 64 Santiago, Chile, March, 2003

Henrion M et al., 2012. Enhancing Access to Electricity in Nigeria through Low-carbon Solutions; Lumina Decision Systems, Los Gatos, California, USA & The World Bank, Washington DC, Triple E Systems, Laurel, MD, USA and Lagos, Nigeria

Igbinovia S O & Orukpe P. E.,2008. Rural electrification: the propelling force for rural development of Edo State, Nigeria; Electrical and Electronic Engineering Department, University of Benin, Nigeria & Control & Power Research Group, Department of Electrical Engineering, Imperial College, London

Iwayemi A., 2008. Investment in Electricity Generation and Transmission in Nigeria: Issues and Options: International Association for Energy Economics

Kennedy-Darling J. et al., 2008. The Energy Crisis of Nigeria; an Overview and Implications for the Future: The University of Chicago

Makwe, J., (2011).The Nigerian electricity Market: University of Aberdeen Business School

Ministry of Mines and Steal Development Nigeria
http://mmsd.gov.ng/solid_minerals_sector/Coal.asp: Accessed 21.04.13

Mshelia H. I.2012., Energy Access for All: The role of clean energy in alleviating energy poverty

Okafor E.N.C. and Uzuegbu J., (2010). Challenges to development of renewable energy for electricity power sector in Nigeria. International journal of academic research, 2:2 211- 216

Ogungbuyi et al., 2012 e-Waste Country Assessment Nigeria: e-Waste Africa project of the Secretariat of the Basel Convention: Basel Convention Coordinating Centre, Nigeria and Swiss Federal Laboratories for Materials Science and Technology (Empa), Switzerland

Oyedepo S.O., 2012, Energy and sustainable development in Nigeria: the way forward; Energy, Sustainability and Society A springer open journal

Okoro & Chikuni, 2007., Power sector reforms in Nigeria: opportunities and challenges: University of Nigeria, Nsukka, Nigeria & Polytechnic of Namibia, Windhoek, Namibia: Journal of Energy in Southern Africa. Vol 18 No 3. August 2007

Onyema Mac-Anthony C., 2010; Alternative Energy Sources for Agricultural Production and Processing in Nigeria: Nigeria Strategy Support Programm Policy Note No. 24

PHCN (2008) Generation and Transmission Grid Operations; Annual Technical Report National Control Center (NCC), Osogbo

Renewable Energy Country Profile: Nigeria: IRENA: The International Renewable Energy Agency

Rilwanua L., (2003). Energy Commission of Nigeria "National Energy Policy"Federal Republic of Nigeria.

Sambo, A. S., (2009). Strategic Developments in Renewable Energy in Nigeria. International Association for Energy Economics. Third Quarter: 15 – 19.

Sorensen. B., (2000). Renewable Energy: Its Physics, Engineering, Environmental Impacts, Economics and Planning. 2nd edition. Academic Press

Sambo A.S., 2006. Renewable Energy Agency Electrification in Nigeria: The Way Forward; Energy Commission of Nigeria, Abuja: Paper presented at the Renewable Electricity Policy Conference held at Shehu MusaYar'adua Centre, Abuja, 11-12 December 2006 Nigeria

Sophie Hebden 2006-06-22 "Invest in clean technology says IEAB report" http://www.scidev.net/en/news/invest-in-clean-technology-says-iea-report.html

Sylvester A, U., & Rabiu I, S., 2012. Electricity Access in Nigeria: Is Off-Grid Electrification Using Solar Photovoltaic Panels Economically Viable? A Sustainability, Policy, and Innovative Development Research (SPIDeR) Solutions Nigeria Project.

Trading Economics 2013 http://www.tradingeconomics.com/nigeria/combustible-renewables-and-waste- percent-of-total-energy-wb-data.html: Accessed 25.04.13

Uzoma, C. C. et al., (2011). Renewable Energy Penetration in Nigeria: A Study of The South-East Zone. Continental J. Environmental Sciences 5 (1): 1 - 5, ISSN: 2141 - 4084

Uyigue et al., 2009. Energy Efficiency Survey in Nigeria; A Guide for Developing Policy and Legislation: Community Research and Development centre

Wagner C., (2006). Modern Wind Energy Plant in Rural Scenary

Internet Sources

Amnesty International 2009
http://www.amnesty.org/en/news-and-updates/news/oil-industry-has-
    brought-poverty-and-pollution-to-niger-delta-20090630

Business Day Nigeria 2013
http://businessdaynigeria.com/nigeria-needs-25bn-investment-gas-
    sustainable-power-reform

IAEEA.Org.
http://www.iaea.org/Publications/Magazines/Bulletin/Bull092/09204700514.
    pdf
IEA (2002). "Energy and poverty." IAEA Bulletin, Paris: International
Energy and Democratic Agency, pp. 24-29

Heather Murdock 2013: General Electric to invest $1 billion in Nigeria
http://www.voanews.com/content/general_electric_to_invest_one_billion_do
    llars_in_nigeria/1594722.html

'Twin Pillars' in Reducing Carbon Emissions: Energy Efficiency Investments
    and Renewable Energy Purchases Together Are.
http://www.aceee.org/press/2007/05/energy-efficiency-investments-and-
    renewable-energy-purch

Solar Power
http://www.solarpower.co.za/

Ministry of Environment Nigeria
http://environment.gov.ng/special-units/renewable-energy/

Mbendi 2013,
http://www.mbendi.com/indy/powr/af/ng/p0005.htm

New statesman 2013: PHCN Index Sores
http://www.newstatesman.com/company-profiles/energy/power/power-
    holding-company-of-nigeria

World Energy Outlook
http://www.worldenergyoutlook.org/resources/energydevelopment/globalst
    atusofmodernenergyaccess/#d.en.8609

http://www.thisdaylive.com/articles/fg-targets-75-electrification-across-
    nigeria/128452/
http://www.sungas.iied.org/challenge

http://www.utwente.nl/mb/cstm/research/urbanenergy/reports/nigeria_rep.doc/
http://livingearth.org.uk/projects/gas-to-power/
http://www.worldenergy.org/data/sustainability-index/country/nigeria
http://www.iied.org/improving-peoples-access-sustainable-energy
http://www.energsustainsoc.com/content/2/1/15
http://www.helio-international.org/Nigeria.En.pdf
http://www.vanguardngr.com/2013/02/the-challenges-of-the-nigerian-electric-power-sector-reform-1/
http://www.informafrica.com/business-africa/interview-the-future-is-bright-for-nigeria/
http://shrinkthatfootprint.com/average-household-electricity-consumption